# 100

## GREEN SCHOOLING

# IDEAS

Towards a Sustainable Culture in Schools

I0479625

## Dr. Dheeraj Mehrotra

INDIA · SINGAPORE · MALAYSIA

# Notion Press

Old No. 38, New No. 6
McNichols Road, Chetpet
Chennai - 600 031

First Published by Notion Press 2020
Copyright © Dr. Dheeraj Mehrotra 2020
All Rights Reserved.

ISBN 978-1-64828-733-6

# Contents

# Preface

Innovation in schools appear a common sight towards excellence for the students, What is required is an awareness to the GREEN Initiative in particular to make a sustainable future for our generations to come. Environmental sustainability doesn't mean living without luxuries but rather being aware of your resource consumption and reducing unnecessary waste. Hence is the priority towards initiating some GREEN SCHOOLING PROJECTS as a common practice.

A Green School is a school that creates a healthy environment conducive to learning, while saving energy, environmental resources, and money. Green schools significantly lower greenhouse gas emissions and energy costs, improve student and teacher health, and enhance student motivation.

Hence this book *"100 GREEN SCHOOLING IDEAS"* is a module which reserves the spectrum for all Quality Schools to practice and implement the environment that is conducive to learning at large through GREEN Initiatives. This shall prove as a ready reckoner for Teachers, Parents and students at large and will further help and inspire the STUDENT fraternity of all age groups to explore the GO GREEN Practices making a delight and  Quality as a way of life.

Don't forget to leave a feedback at tqmhead@aol.com

**– Dr. Dheeraj Mehrotra**
Academic Evangelist,
TEDx Speaker & National Awardee
www.dheerajmehrotra.com

"The planet is, was, and always will be stronger than us. We can't destroy it; if we overstep the mark, the planet will simply erase us from its surface and carry on existing. Why don't they start talking about not letting the planet destroy us?"

– Paulo Coelho

## GREEN INITIATIVE #1

Consider Going Solar. It has to be our objective to promote the Environmental Stewardship through both classroom and outdoor activities like Teachers Training Workshops on digital Pedagogy, development of green curricula in particular.

Improve and Implement Recycling Programs. This may include community impact projects, outdoor educational activities, development of information education and communication resources for joyful learning activities on priority.

Improve Your Air Quality

Avoid burning of fossils. Practice the
first law of Environmental Education: An
Experience is worth 10,000 pictures.
"The ecologization of politics requires us to
acknowledge the priority of universal human
values and make ecology part of education
and instruction at an early age, molding a
new, modern approach to nature and at the
same time, give back to [us] a sense of being
part of nature. No moral improvement of
society is possible without that."
Mikhail Gorbachev

Execute Nature Rituals & Co-curricular Activities. Take part in role plays and simulations or get involved in a community outreach project involving an environmental issue like NUKKAR NATAK (Street Play). We can motivate the students to plant trees, plan an environmental running game or perform an environmental dance item.

Get Plants for the Classrooms. The best plants include Areca Palm, Raphis Palm, Bamboo Palm, Rubber Plant, English Ivy, Pothos, Ficus Alii. The Snake Plant, Spider Plant, Garden Mum, Boston Fern and Even Peace Lily all are low maintenance plants and help in purifying the air in the classroom/home/offices.

Prefer Clean and Green as a priority within the campus. Encourage Parents to use public transportation. Encourage your school students to sing the Green School Initiative as a document. Maintaining cleanliness in the school helps reduce student absences and teacher sick days. The research shows that School Cleanliness affects Students Performance.

GREEN INITIATIVE #7

$U$se GREEN Building Materials. This may include natural materials like cork and bamboo to recycled glass. The recycled metals reduce the energy consumption and air pollution and decreases the over harvesting of trees, landfill waste and natural ore mining. Cork allows excellent thermal insulation, non-slip resistance, it is hard wearing and safe making it the perfect choice for flooring and insulation and increases the energy efficiency in the process.

## GREEN INITIATIVE #8

Use Green Cleaning Products. Most green cleaning products also come in refillable containers, which mean less plastic use. It smells more natural, free from harsh chemicals. This includes the Disinfectant Cleaners, bio clean, Enviro Care Neutral Disinfectants. This reduces carbon emissions and boosts test scores, protects the health of the custodial staff, preserves the environment and helps students stay healthy.

Reduce Your Energy Use in the schools. Turn off the appliances when not in use, use the building systems properly, upgrade the heating controls, appreciate using both sides of the paper. Sell or donate goods instead of throwing them out. Consider reusable products. Borrow, rent or share items.

Test Your Water and review the findings. Lead level in water may cause damage to the brain and the nervous system, it slows the growth and development, gives rise to leaning, behaviour, hearing and speech problems for the students. It is recommended to flush pipes before use and avoid consuming hot water from the tap, install a lead filtration system and replace the lead pipes and install a lead filtration system in place.

Encourage Pupils To Walk to raise environmental awareness. It also reduces congestion around school, improves the social aspect of coming to schools as children, parents and teachers get to chat on the way and promote an awareness towards global environment. It can take the brand of a "Free Your Feet" initiative.

Eat Green using the Garden to Café Program. The schools may support the food grown in school gardens directly to the students. The menu may be designed with seasonality in mind and farm fresh initiatives. This way they also know how to grow nutritious food crops as well as how to cook, share and enjoy them at meal hours.

Explore RECYCLE

As a hobby in the school by adding into the mix, right sizing dumpsters and collection. It shall reduce the fossil fuel usage, conserve resources. The module works well by creating a classroom level audit for waste. Each class must have its own dustbins. The class representatives need to effectively manage the waste under the guidance of the teachers.

## GREEN INITIATIVE #14

Distribute Rewards for Cleanliness and Recycling. The students may be honoured or praised in public for the good job done. This shall definitely encourage them to recycle and shall include more and more students in a waste recycling programme. Teachers may also be included in the recycling programmes as an encouragement to the waste recycling and the protection of the environment.

Make the OLD into something new to bring attraction to the Recycling process by creating animal characters to teach students what they can do to help our planet. Promote Recycle of cartons in the school and in the neighbourhood. Even students need to be made aware up to 80 percent of a phone is recyclable, so do not make it a treasure instead recycle it!

Encourage Healthy Living and Eating Practices. Plan healthy meals and eat together as a family. Go on more fruits and vegetables : fresh, frozen or canned, get fewer soft drinks and high fat/ high calorie snack foods like chips, cookies and candies. Schools must focus on overall diet rather than specific foods. Get kids involved in shopping for groceries and preparing meals.

Be Energy Efficient, make the students use less oil, coal and fossil fuels. Reduce the campus' energy consumption which shall help to reduce the air pollution and solid waste.

Take advantage of the natural sunlight, switch over to LEDs or CFLs and invest in energy-saving power strips. Insulate the roof spaces and install timed tap systems.

Move to Sustainable Classrooms by turning lights off in every empty room no matter the time of day. Let in more sunlight, organise monthly "lights off hour", Walk the talk, with sustainable practice in place of reduce, reuse and recycle. It aims at practicing something that maintains a condition without harming the environment.

Encourage Learning From the Environment. This way the students engage and interact to learn new skills. The concept is to learn from the environment away from the traditional classrooms. The Green Environment based learning offer healthier classrooms and boost the test scores. This mode of learning stimulates the imagination, creativity and innovative learning.

Get Involved in the Community. It provides friendship, warmth, unity, sense of satisfaction, connection and strength. It provides a profound impact on the students and their wellbeing. Social interaction is good for kids. This may happen over luncheons, meeting like-minded people always bring delight through diversity and sharing the issues that are important.

Bring Back and Encourage Fresh Air & Daylight. For sure fresh air increases student performance. Too much of $CO_2$, fine dust or vapours emanating from building materials and furniture result in muggy atmosphere within minutes. And hence we require extensive organisational, ventilation and structural measures on priority. Install convenient and efficient air diffusers and high precision components for air distribution towards achieving an effective air change and optimum air supply within classrooms.

Activate Classroom Cleaning Periods on the Go! We welcome the reads on our social media platforms about schools leading this initiative in Japan, for sure, this includes serving lunch duties and cleaning tables, the classroom cleaning period and the brief recess following the cleaning.

Implement Pest Control and Maintenance Policies as a safety and security Measure for schools. The school should take in view in practice the environmental friendly treatments to get rid of pests as it is safe and perfect for schools. Insecticides used should be eco-friendly.

Work for Reserve Funds for Quality of Air Meters at place. This is one of the ways to educate students about different types of air pollutants like ozone, nitrogen dioxide and how they impact them and the environment in particular. A good Indoor Air Quality is very important. Poor indoor air leads to reduction in the ability to concentrate, lean the leads to discomfort in breathing.

Grow Plants as a Birthday Commitment by Students. Schools need to promote planting saplings by the students on their birthdays to give birth to a green future. Schools may launch 'Plant a sapling on your birthday'. The ministry of Human Resource Development (MHRD) has issued instructions to all the government schools, upto class 8, to include the produce from the kitchen gardens in the mid-day meal! Let us follow the order religiously.

Plan and Execute a Local Food Hour. Every moment of food sharing has to be checked and monitored. The school nutrition environment and services need to be a part of the school learning outcome to model and help students to shape lifelong healthy eating behaviours. The schools must implement policies and practices to create a nutrition environment with fruit and vegetable consumption on priority.

Encourage Vegetable Meals as schools are key environment to teach children healthy eating. There has to be an encouragement towards healthy eating and help them being positive towards their dietary behaviours. Fruits and vegetables need to be promoted for all meals. Serving whole grains, fruit and vegetables has to be a priority.

GREEN INITIATIVE #28

Organise Green Clubs and Events.
Eco clubs in schools tend to empower
kids to participate and take up meaningful
environmental activities and projects.
Starting an environmental club is a great way
to get students energized about taking care
of the earth and helping their community of
the most important issue of climate change.

Encourage Carpooling. It is like sharing of car journeys so that more than one person travels in a car and saves fuel and traffic on the road. Students may be motivated to have carpooling, rideshare, preferred parking for carpools and ultimately saving money and benefiting the environment at the same time. Carpooling makes sense financially and carbon emissions.

Encourage Cycling. This reduces the emissions which therefore reduces damage to the environment in particular. It also reduces congestion directly around the school. It encourages independence and builds confidence among the kids. It also provides an active start to the day and the kids become more alert in their lessons.

Encourage Walking among the kids. Invite them to walk with you. Research a safe walking route for the school campus. Do a practice walk. Get a walking buddy. Start a walking club. Promoting walking to school enhances psychological well-being, self-esteem, social and moral development, reduces over weight and obesity and instils the habit of integrating walking into their everyday lifestyle.

## GREEN INITIATIVE #32

Make use of Glass Water Bottle and Avoid Plastic Water Bottles. In a lot of places bamboo products are replacing plastic. An eco-friendly water bottle has been developed by former IITian from Assam, Dhritiman Bora, the bottle is made of completely natural products and is sealed using a cork which makes it leak proof, so there is no tension there.

Plan a Drinking Water Bell. The school authorities have to initiate a 'Water Bell' which would aim at encouraging students to drink a lot of water during the day to stay hydrated and fit. This can happen TWO to THREE times during the school hours to remind student to pick up their bottles and drink some water.

Encourage Installation of Energy Meters. This is to monitor the use of energy. Affects reduction in energy uses and saves on money, carbon emissions which further helps combat climate change. Schools may calculate their carbon footprint and accordingly provide an opportunity for pupils to be involved in responsible use of energy and water.

Post and Showcase Conversation Reminders and Quotations Around the Campus to promote Awareness. Introduce 3 R's: reduce waste, reuse resources and recycle materials. Organise tree planting days at school and encourage children to switch off all appliances and lights when not in use.

Provide Recycling Bins Around the Campus. Choose your recycling bins based on capacity, location, flexibility and customize as per the need. Plan the pick-up schedule. Communicate and educate the initiative to all the stakeholders. Make sure the bins are clearly labelled with details about what does and doesn't go it them.

Encourage Community Clean-up Efforts through active participation of the Parents/ Students/ Teachers. This helps in building environmental stewardship and teach students the importance of reducing, reusing and recycling to decrease the amount of waste we produce.

Plan and Execute a School Garden and Donate Seeds as Return Gifts. Having a school garden allows students to learn focus and patience, cooperation, team work and social skills. The students also gain self-confidence and a sense of capableness. This gives a new and sustainable perspective on healthy eating habits.

Maximise the alternative ways to produce energy. Work on Solar Energy as a Practice. We can take advantage of the natural sunlight, switch over to LEDs or CFLs, invest in energy-saving power strips, replace TVs with flat and LCD screens. Invest in cooling options, encourage students to recycle, use sensors for turning lights on or off in a room.

## GREEN INITIATIVE #40

Minimise the use of PAPER in the school. Go Paperless. Use computers/Printers/copiers whenever possible. Let the students submit their papers and other homework via email. Even teachers can put all assignments and handouts using a blog or a website.

Use both sides of the paper whenever required. The schools should buy recycles or alternative paper products. Assure the school uses napkins, paper towels and bathroom tissue made from recycled paper and post **"THESE COME FROME TREES"** to promote its limited use among the students.

Form a Green School Alliance and share the initiative through exchange programmes and workshops. This allows children to get set to work on cultivating flower and vegetable gardens. This in a way become a wonderful way to use the schoolyard as a classroom, reconnects the students with the natural world. The children who garden get excited about tasting the fruits of their labors. It also is a powerful environmental education tool.

## GREEN INITIATIVE #42

$E$ncourage the students to use public transportation. This way they become more independent and responsible at an early age than their peers. It is an environmentally friendly mode of getting around. It also allows space-saving nature of public transportation which may help enable denser land use, parks and nature preserves.

Encourage the Students to wear recycled shirts/ T-shirts made from plastic bottles. Plastindia Foundation, the apex body of plastics industry in India, said it look up the initiatives to create awareness of the need to manage plastics waste and recycle them more efficiently. It said it wanted to achieve sustainability through the optimal use of plastic waste.

Reduce idling outside the school. Ask the parents/ bus drivers to turn off car and bus engines when parked. This way there is a reduction in the amount of air pollution that is emitted, reduced health impacts and it also saves money. It is reported that the school buses typically burn up to half a gallon of diesel fuel for each hour that idle.

Encourage Parents and School Buses to use Biodiesel or Electric Options to Fuels. The biodiesel refers to a vegetable oil- or animal fat-based diesel fuel. The surge of interest in biodiesels has brought a number of environmental effects associated with its use. There is a potential reduction in the greenhouse gas emission, deforestation, pollution and the rate of biodegradation.

## GREEN INITIATIVE #46

Encourage the Parents to use compact fluorescents instead of existing incandescent light bulbs. CFLs use 50–80 percent less energy than incandescent lights. CFLs are less expensive in the long run as they last much longer than incandescent bulbs. The CFLs are compact and are ideal for use in homes, work areas, schools and at workshops.

Use Environmentally friendly cleaning fluids and nontoxic pesticides. There has to be a framework for eliminating the underlying conditions to allow pests to thrive and for using least toxic methods like baits and traps to avoid unnecessary pesticide exposures. This leads to a safer and healthier environment for the students.

Encourage the students to avoid using plastics as Tiffins. The schools need to ensure that students no longer carry plastic tiffin box and water bottles to schools. The fact is these plastic tiffin boxes wear away after a few washes, they also contain plastic which may constitute a harmful chemical called Bisphenol-A (BPA).

## GREEN INITIATIVE #49

$E$ncourage the Students to reduce the Water Usage. There has to be some initiatives in schools to draw everyone's attention to the importance of universal access to clean water, sanitation and hygiene facilities. A talk on water conservation may be held during the morning assembly.

Make use of Reusable Bag Practices as a priority. Schools must motivate students to have reusable grocery bags which is a structured tote bag, ideal for packing the lunch, carry snacks, take books from the library in addition. It is an alternative to single-use paper or plastic bags. In some cases, these may be used over 100 times and are better for the environment than the single use plastic bags.

## GREEN INITIATIVE #51

Encourage the Students to Sign
The green school pledge. This in a way can
have a significant impact on the environment.
This way the students are encouraged
to help their families save energy, waste
and water at home by implementing the
commitments on the pledge in order.
This enriches cultivation of a healthier and
responsible school environment.

## GREEN INITIATIVE #52

Encourage the students to AVOID using LIGHTS to save on energy during the day time. It is observed that plugged in electronics are energy vampires even when powered down. Students must practice simple habits leading to big savings both for the environment and for the school budget. For sure, switching off the lights when leaving the room is a wonderful start for itself.

Adopt An Environmental Vision Statement of the School. This may include invitation for a community atmosphere and beautiful surrounds in a positive learning environment. The mission should be to advocate for the increased practice of environmental education, promote the development and use of effective, research based environment education pedagogy. The objective must be to support the values of diversity in environmental education in particular.

$O$rganize an Academic Audit with emphasis on Environmental Survey. A green school audit is a tool drafted to assist schools to audit their use of natural resources and provides them the opportunity to become environmental managers by assessing themselves. The priority includes the setting up of the audit teams which would assess the schools environmental practices. It helps the evaluation of the use of resources and map their consumption and wastage.

Adopt a GREEN School Action Plan. This has to be documented and should involve and promote the action as a school's recycling program, the priorities include being eco-friendly, usage of non-toxic cleaning materials, carpooling as a practice, energy conservation like switching off lights when not in use and others as monitored.

Adopt a GREEN SCHOOL curriculum has to have a vison of a natural, holistic, student centered play based learning environment that empowers and inspires the students to be creative and innovative. It has to create a strong sense of community, emphasising the relationship between the students and the teachers through increased wellbeing and less fear. The objective has to be to learn how to lean as a life long learner.

Organise Awareness Walkathon to spread Green Initiatives. This may include a green race at school to protect the environment. The students may even plat saplings to finish a race and put vegetables in a cloth bag in a hurdle race. Also during the walkathon the students may carry posters to showcase the "AIR PURIFICATION" at homes as a priority, Avoiding plastic bags, among others.

## GREEN INITIATIVE #58

Start a Student Run Recycling Club under the Supervision of the House Incharges. This may assist schools in implementing, maintaining and improving recycling programs. It is estimated that 24% of the school waste is recyclable paper and 50% is food waste and non-recyclable paper that can be composted. Establishing a recycling club would encourage the recycling throughout the school.

## GREEN INITIATIVE #59

Make use of Recycle Newspapers and Magazines to Create Art Projects. The children can be guided to craft ideas from jewellery to wall décor. One may create fun embellishments, a plant holder for the living rooms, turn the comics into beads of jewellery, the sports section into faux flowers and start some seedlings growing in the home and garden section, funny flowers through comics from newspapers, comic wrapped decorative balls, upcycled newspaper earrings and other wonderful items.

Host a Solar Powered Cooking With Parents as Volunteers. Promote the cause as an Annual Event Class wise. The cooking needs to be done on a sunny day and is well suited for a school fete or a green day. They use a free, renewable energy source and do not pollute the environment. In addition, projects may include solar energy, solar power, effects of heat, recycling, living green, cooking without/with fire.

Organise an Environmental Informative Field Visit for the Students. It can be referred to as a Flora & Fauna Field Trip as an Annual Event class wise. This can be one of the most popular ways for introducing students to the concepts, ideas and experiences that may not be provided within classrooms. This may lead to museums, zoos, aquaria and outdoor study centres during the academic year.

Create a Bird house Habitat Around the playground or in the campus around to spread awareness about the importance of sparrows and other endangered birds. The students may use cardboard and discarded papers to decorate the birdhouses. This is one of the ways to create awareness among the students and also encourages them to take up initiatives to work for the society in future.

Adopt an Endangered Animal in the Zoo. The adoption programme enables schools to participate and make a valuable contribution towards the care and enrichment of the animals. It also strengthens the zoo's wildlife conservation programme.

## GREEN INITIATIVE #64

Promote Taking Notes Electronically and avoid use of PAPERS. Use computers whenever possible. Submission of projects and assignments may be active online as a practice. If required, use recycled paper and work on both the sides. Also in process the schools must educate staff and students about the environmental and economic impact of paper and ways to conserve.

Adopt a Planet Pledge for all the Students in the school. Take the Planet Pledge. This may include loving the planet, being proud of its rich mineral resources, planting more trees and keeping the surroundings clean including usage of environmental friendly products as a habit.

Make use of Old Water Bottles to quench their thirst. Prefer using glass water bottles and avoid using plastic water bottles. To much a say, the education department has banned the use of plastic water bottles in government schools on the eve of the World Environment Day in Haryana, India. Instead the circular quoted, water bottles of alternative material such as glass, steel, aluminium can be used.

Work on decreasing the MEAT consumption. Eating less meant-even omitting it from a meal once day a week can influence change. There is a need to make a meat-free day with meals based around plant-based proteins, such as beans and pulses, compulsory each week. Meatless Mondays are being introduced in global school canteens.

## GREEN INITIATIVE #68

Generate awareness towards avoiding Aerosol Sprays as beauty products, instead use environmentally responsible brand seals. Make awareness towards skipping products that are sprayed directly on the body like sunscreen, hair spray and deodorant. Choose non-spray options instead.

Practice COMPOSTING as a Nature's Recycling Plan. This helps in turning organic waste and reduces trash going to the landfill. It provides a useful way for teachers, pupils and catering staff to explore educational opportunities together. Classroom composting using worms need to be more common. The finished compost can be used in the school gardens.

$M$onitor and Measure the $CO_2$, Temperature and Humidity Levels in the Classrooms and the AIR INDEX with emphasis on Water Purity. Indoor air quality in schools has received a great concern over the years. When the school air quality is unhealthy, learners suffer. It leads to decreased concentration and poor test results.

Promote the use of Sustainable Supplies. Make use of paper products made with recycles content. Plastic Products should be labelled as PVC free. Identify the Energy Star Logo and rating on the computers and other electronic devices in use.

Develop a Timeline for Completing the Green Projects, prepare a Milestone Chart, Appoint Green Team Members and Mentors. This leads schools to set up and run a student-led environmental programme. The timeline may include fun and engaging environmental projects, from recycling clubs to solar cookouts.

Make a check on water dripping which is a common thing around the school campus. Check that the water faucets are turned off. Tap in charges need to report the water dripping to the student council. Identification of leaking taps at schools. The schools need to have a check on how much water does a leaking faucet waste as a priority.

## GREEN INITIATIVE #74

Invite Consultants as panellists during school functions to identify and work towards Energy Efficiency through students. Plan out on strategies to save energy. The consultants need to deliver workshops/awareness camps and empower students to have energy conservation projects.

Calculate the Carbon Footprint or the combined effect all the students have on the environment. This may lead to further projects on steps to reduce the same. Options may include going with greener alternatives, encourage active walking, reduce energy waste, use air conditioner and heater wisely. In addition compost all non-animal based organic materials.

Make use of Recycled Pencils Only. The use of trees for the manufacturing of pencils leads to unnecessary deforestation. Several companies offer pens made from recycled plastic. Pencils can be made from recycled newspaper. We may save one lakh trees by getting schools to use the eco-friendly pencils.

## GREEN INITIATIVE #77

Identify Green Sites within the campus and measure the ecosystem services available for the students and the society. The schools may go for adding solar powered tables to hot spots on the campus. This project promotes environmental stewardship in schools through classroom and outdoor activities to improve critical, interdisciplinary and holistic thinking.

## GREEN INITIATIVE #78

$O$rganise Online Classes as a periodic practice which may save travel resources as well as the energy costs of the brick and the mortar. Of course, online learning reduces the negative environmental impacts that come from transport and construction. The cost to textbooks, desks, tables, electricity, buildings, ground are all reduced which further reduces waste and conserves the natural resources in a big way.

Organise a recycling BIN Decorating Contest. This may be an Inter School or an Intra School Event. Students can participate and suggest ideas on single use plastic bags alternatives which are cost effective, use natural textiles with robust designs for groceries, vegetables and family bags.

Make use of Rechargeable Batteries which can save the earth from harmful metals and compounds that can't be broken down otherwise in case of normal one time use batteries. The rapid progress of PEDs (Portable Electronic Devices), have also lead towards the demands for innovative rechargeable batteries technology.

The leftover drinking water from the water bottles may be used for storage to water the plants, when the children are about to leave for home. They may even collect the leftover water from their water bottles in a large drum installed in their school. This water replaces the fresh water used to water the plants and clean the school premises.

The Schools Need to have a Cloud Based Computing and Data Storage to avoid Paper Usage. Schools can greatly benefit from adopting cloud computing. They actually reduce costs, promote a better learning environment for the students and create a better learning environment for the students in particular. Students can even appear in virtual exams saving lots and lots of paper, writing time and furniture.

Run a Green Campaign. Make green living a part of every day school life by considering the environment in everything we do. The objective of this has to be to educate the children on their responsibility towards the environment and also to make them involved and engaged in the relevant projects on the environment. This would lead initiatives towards Save the Planet as the chief motto in particular.

Explore a habit of Packing and Recommending a NO-WASTE lunch. There has to be no leftovers in food. Make use of sustainable packaging like reusable cloth bags, stainless steel, glass containers, cloth napkins and reusable tableware in particular. Some of the zero waste lunch ideas include sandwiches, wraps, pasta with pesto, green salad with chopped vegetables among others. Reduce use of plastic water bottles and waxed cardboard drink boxes by packing a healthy drink in an insulated glass mug or metal water bottle.

Play movies/youtube videos with Environmental themes. Recite poems on Sustainability and Sing Songs on Environmental Issues/Protection/Pollution Control on the go. The environmental awareness and reporting on environment and development has come of age in India. TV is an excellent medium for promoting environment awareness, development, literacy, social awareness and pollution control measures.

## GREEN INITIATIVE #86

Encourage kids to buy GREENER FABRICS, whenever possible, opt for planet-friendly fabrics like organic cotton or hemp, repurposed or recycled materials.

$S$et up WORM BINS. As we know Worms love turning food waste into gorgeously perfect compost. Even the harvesting of rainwater from all the school buildings is a possibility. The dirty water from the kitchen can be recycled and sent to the organic garden.

Organise a Tree Plantation drive as a commitment to conserve the precious fauna. It connects the children with nature and shows them that their small contribution makes a big difference to the campus making it greener. Schools may plant different species or rare trees and plants. The initiative to have classes named after the trees planted in the grounds such as Ashoka, Champa, Chameli, Arjuna and Krishna act as a wisdom means of sustaining the environment to be green as a hobby in process.

## GREEN INITIATIVE #89

Encourage interaction on environmental issues, get students involved in surveying the class on point of views and defend their reasoning via discussions and live interactions. Start a "Questions to Investigate" corner or a "Problem of the Week" competion. This way the students have an opportunity to compare and contrast, even explore biases and try to understand the complexity of the burning environmental issues in particular.

## GREEN INITIATIVE #90

Prescribe buying USED TEXTBOOKS. Available for half off or more on websites like Amazon. Renting or buying used textbooks is an increasingly popular option which helps reduce the creation of books and hence can save millions of trees.

Every School Must act as a LIGHT HOUSE for the society towards Green Education. Be an enviroschool and bring sustainability education to the students through early education and empower them to think and act sustainably.

Participate in National and International Award Programmes that guides schools through a process to help them address a variety of environmental issues, ranging from litter and waste to healthy living and biodiversity.

Make a Green Club with projects and activities. Organise one GREEN PROJECT deliberation as awareness every week during the morning assembly.

Plant a RAIN GARDEN. It is a garden filled with native perennials designed to capture runoff rainwater and recycle it back into the ground to reduce the pollution and preserve the sewer systems.

Organise a recycling BIN Decorating Contest, may be Inter School or Intra School to generate the awareness of GREEN Initiatives to save the environment. Schools may take it as an Earth Day project so as to make it fun for the family too.

Rally for SOLAR panels. As we know Solar is the cheapest energy source and saves a ton of money. Get Set Go. Make it mandatory for schools and homes to have at least one solar panel in every session. Add the tally to make it more sustainable. Beyond doubt, the solar power is helping schools all over the world to conserve energy costs and help communities to be more environmentally friendly.

Schedule a TRASH Pickup TIME TABLE hour. This may be once a week or once in 15 days class wise. Accordingly the pick up of trash, recycling and green waste carts should be placed along for the scheduled pick up day. The objective also has to be to reduce the schools' environmental footprint.

Organise SPECIAL Assemblies/ Panel Discussions/ Workshops for Students/ Teachers/ Parents to understand WHY recycling is important as a regular event in the school. The activities should be able to raise awareness among students to save environment delivering the fact "A better environment means a better tomorrow."

## GREEN INITIATIVE #99

Make use of Acid-free Glue Stick for all the art projects. It is less messier than liquid glue and better for the environment in particular, as they are made from 90 percent renewable ingredients. They are environmental friendly and safe chemicals, bio-degradable and can be washed out.

Install a SENSOR based lighting System as one of the easiest ways to save energy at school. Even a modern school lighting can be designed specifically for classrooms, with standards to minimise power consumption and $CO_2$ emission. It delivers the greatest potential for energy savings in school buildings.

# About the Author

Youth Ki Awaaz @YouthKiAwaaz · 17m

Meet @DheerajMehrotra, the record breaking teacher whose 150+ apps are changing how we learn: yka.be/2i9FKxB

**Dheeraj Mehrotra,** MS, MPhil, Ph.D. (Education Management) honoris causa., a white and a yellow belt in SIX SIGMA, a Certified NLP Business Diploma holder, is an Educational Innovator, Author, with expertise in Six Sigma In Education, Academic Audits, Neuro Linguistic Programming (NLP), Total Quality Management In Education, an Experiential Educator, a CBSE Resource towards School Assessment (SQAA), CCE, JIT, Five S and KAIZEN. He has authored over **40** books on Computer Science for ICSE/ ISC/ CBSE Students, over **10** books of academic interest for the field of education excellence and Six Sigma. A former Principal at De Indian Public School, New Delhi, (INDIA) with an ample teaching experience of over Two Decades, he is a certified Trainer for Quality Circles/TQM in Education and QCI Standards for School Accreditation/ Six Sigma in Education. He has also been honored with the **President of India's National Teacher Award in the year 2006** and the **Best Science Teacher State Award** (By the Ministry of Science and Technology, State of UP), **Innovation in Education** for his inception of Six Sigma In Education by Education Watch, New Delhi and **Education World-Best Teacher Award,** BOLT Learner Teacher Award by Air India, **'Innovation in Education Award 2016'** by Higher Education Forum (HEF), Gujarat Chapter, among others. He has developed over **150 FREE EDUCATIONAL MOBILE Apps** for the Google Play Store exclusively for Teachers, Students and Parents. This work has been recognized by the **LIMCA BOOK OF RECORDS & INDIA BOOK OF RECORDS** as the only Indian to draw that feast. Dr. Mehrotra is presently working as an **Academic Evangelist** in India. As a UDEMY Premium Instructor, he has over 250 Online Courses with around 5 Lakhs enrollments from 180 plus nations around the world. He is an active TEDx speaker and can be viewed at https://goo.gl/m86TUf

# Notes

..............................................................
..............................................................

..............................................................
..............................................................

..............................................................
..............................................................

..............................................................
..............................................................

..............................................................
..............................................................

..............................................................
..............................................................

..............................................................
..............................................................

*Notes*

_Notes_

## WORLD EDUCATION BROADCAST

EDU TV is a progressive education channel dedicated to bring the latest happenings in the world of academics and education to its audience. We have a wide coverage of the latest events, discussions, and who's who and what they have to say about the evolving education sector. It showcases the unprecedented work of 'The Knowledge Confluence' which has been relentlessly working for over a decade now, in bridging the lacuna between what the aspiring students are seeking and the national and international Universities have to offer. The goal of connecting students and their families with the multiple choices and wide range of available options in higher education so that they make an informed decision is met year after year by connecting the Universities and the Students at fairs and events organized by The Knowledge Confluence. It covers Principal's Conclaves where the eminent academicians exchange best practices and discuss the evolving trends and challenges of the education system

## Books by the Same Author

Available at amazon.com

# Information on Green Schooling Certification

Exceptional Learning & Limitless Opportunities...

EARN A GLOBAL TEACHING DEGREE

**Full Time & Online**

This Course is designed to be completed in one year containing various Nature Skills that empower you to bring power of nature into all State, National and International Curriculum.

The Topic incorporated into its curriculum are simple and easy to understand. If you have interest in the teaching field, then you will be able to finish the program smoothly.

Graduate aspiring to build their career in Green Teaching should possess these traits and skill sets i.e. Passion for Environment, Passion for Teaching, Patience, Discipline, Enthusiasm, Communication skills, Creativity and Management skills.

Diploma in Green Teaching opens ocean of opportunities in Green Schools across the World, as Green Teacher, Principal, Administrator, Green Auditor, Green Mentor, Trainer, Writer and Content developer.

JOY OF TEACHING EVERY DAY AND EVERY WAY

118